The Big Trip

written by Carrie Waters
illustrated by Rick Brown

SAXON
PUBLISHERS

THIS BOOK IS THE PROPERTY OF:

STATE_____ PROVINCE_____ COUNTY_____ PARISH_____ SCHOOL DISTRICT_____ OTHER_____	Book No. _____ Enter information in spaces to the left as instructed

ISSUED TO	Year Used	CONDITION	
		ISSUED	RETURNED

PUPILS to whom this textbook is issued must not write on any page or mark any part of it in any way, consumable textbooks excepted.

1. Teachers should see that the pupil's name is clearly written in ink in the spaces above in every book issued.
2. The following terms should be used in recording the condition of the book: New; Good; Fair; Poor; Bad.

The big trip is on!

Big Bob hops in.

Big Bill sits on top.

Big Babs is last.

Big Babs fills it.

Big Babs tips it.

Bam! Plop! Plop!

The big trip is off!

The End

Understanding the Story

Questions are to be read aloud by a teacher or parent.

1. What is the title of this story?

2. Why do the animals get into the hot air balloon?

3. Why is the big trip off?

Answers: 1. The Big Trip 2. to take a big trip 3. Possible answer: because the balloon cannot fly with all that weight

---

Saxon Publishers, Inc.
Editorial: Barbara Place, Julie Webster, Grey Allman, Elisha Mayer
Production: Angela Johnson, Carrie Brown, Cristi Henderson

Brown Publishing Network, Inc.
Editorial: Marie Brown, Gale Clifford, Maryann Dobeck
Art/Design: Trelawney Goodell, Camille Venti, Andrea Golden
Production: Joseph Hinckley

© Saxon Publishers, Inc., and Lorna Simmons

All rights reserved. No part of the material protected by this copyright may be reproduced or utilized in any form or by any means, in whole or in part, without permission in writing from the copyright owner. Requests for permission should be mailed to: Copyright Permissions, Harcourt Achieve Inc., P.O. Box 27010, Austin, Texas 78755.

Published by Harcourt Achieve Inc.

Saxon is a trademark of Harcourt Achieve Inc.

Printed in the United States of America
ISBN: 1-56577-951-7

4 5 6 7 8   446   12 11 10 09 08 07 06

Saxon Phonics and Spelling K

Phonetic Concepts Practiced

b (big, Bob)

ISBN 1-56577-951-7

Grade K, Decodable Reader 5
First used in Lesson 59